Harriet Cushman Wilkie

Cross-Stitch Embroidery

Harriet Cushman Wilkie

Cross-Stitch Embroidery

ISBN/EAN: 9783337254711

Printed in Europe, USA, Canada, Australia, Japan

Cover: Foto ©berggeist007 / pixelio.de

More available books at **www.hansebooks.com**

The Priscilla Cross-Stitch Book

Price, Twenty-five Cents

Published by

The Priscilla Publishing Company

Boston, Mass.

Cross-Stitch Embroidery

BY

HARRIET CUSHMAN WILKIE

PUBLISHED BY

THE PRISCILLA PUBLISHING COMPANY

BOSTON, MASS.

PRINTED BY THE T. W. RIPLEY CO.
THE SOUTHGATE PRESS, BOSTON

Cross Stitch Embroidery

By Harriet Cushman Wilkie

The familiar cross stitch is at once one of the oldest and most simple of embroidery stitches. There is historical evidence that it was in common use with the Phrygians, the Egyptians and the Hebrews, and it is no stretch of fancy to believe that the rich pomegranates on the linen hangings of the Tabernacle were formed by this stitch. So popular has this style always been, that every decade or two sees a revival. Perhaps the most notable occurred in the 11th century, when many famous tapestries were embroidered in cross stitch.

During the 13th century cross stitch again became the popular style of embroidery, and as it was extensively used by the nuns in the numerous convents for kneeling mats and altar cushions, it was called "cushion style." It was also used for Church vestments. Many elaborate specimens are still preserved.

It was as far back as 1804, that colored patterns were first published in Berlin, and the old style then received a new name, by which it is usually known at present. The method of twisting and dyeing crewels was invented about the same time, and when our grandmothers were maidens, the new-old Berlin work flourished in its glory. The specimens of embroidery of the period seem crude and inharmonious now, for in the present revival of this old method of embroidery the designs, colors and applications are effected by wider knowledge and higher taste in art.

There is much confusion on one point of nomenclature. Russian work may be done in cross stitch, but all cross stitch is not Russian. The distinction lies in the colors.

The real Russian designs are different from the German, and in the latter more variety of color is used, excepting in the work found in the eastern provinces of Russia. Russian designs have a quaint character of their own: in the German, heraldic animals and scroll work appear, and they are altogether more elaborate.

The true Russian work, became first known to any extent in England after the marriage of the Duchess of Edinburgh, and the first specimens of it of any importance which were seen were on the towels forming part of her Royal and Imperial Highness's trousseau, and kindly lent by her to the exhibition of lace held in that year at South Kensington. It may be best defined as embroidery in red and blue cotton upon linen, in conventional patterns, executed in *three* different stitches, only one of which, *point croise*, or cross stitch will be considered here.

Russian work, we need hardly say, is now greatly in request for almost every article on which it can possibly appear, not only on linen for tea cloths, towels, serviettes, sideboard cloths and bedroom furniture, but on silk for borders, ties, gloves and slippers. The linen canvas exactly reproduced from the old material is the best for working upon where it can be used; but when Russian embroidery is applied to fine linen or silk, ordinary canvas must be first tacked on to the material, and the work done over it, the threads to be afterward drawn away. The linen canvas should be worked either with silk or cotton of ingrain colors. A "purist" will, however, say that the Russian work should only be seen in the true Russian colors, the positive red and dark blue, and decided yellow and green, the pale and half tints, light pinks and lilacs being wholly inadmissible. The Russian apron illustrated here is made of strips of twill in red, blue and white worked in cross stitch in embroidery cotton with two shades of blue, one each of red and yellow and white and black knitting silk. The

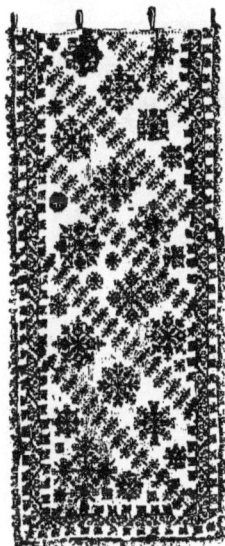

Fig. 1. German Sham Towel, used as a wall decoration

Fig. 2. Russian Apron

blue strips may be worked with red, white and yellow, the red ones with dark blue, white, black and yellow, and the white ones with all the colors.

CROSS STITCH. The first stitch illustrated (Fig. 3) is the familiar cross stitch on canvas, and is so simple that a description seems needless. One point however must be noted, all stitches must be crossed in the same direction, and care must be taken to avoid drawing the needle through the threads of the canvas. The purposes for which this stitch may be used are simply legion.

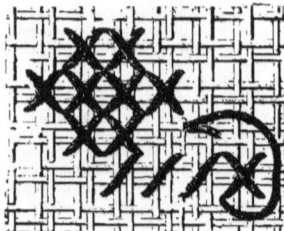

Fig 3   Cross Stitch

The secret of the long-continued success of cross stitch lies in the very fascination of seeing the intricacies of the pattern gradually growing beneath the hand, and in the consciousness that the power of producing any particular pattern lies with the worker herself, and is by no means dependent upon a machine-printed model.

Any cross stitch pattern may be enlarged by counting four, nine or any perfect square for each stitch in the design.

USE OF SYMBOLS. Many of the designs are to be worked in many different colors, each of which is represented in the design by a certain mark or symbol, printed within the square. Thus one stitch may be a cross, another a circle, one white and the next black, or gray. These will then be signs of different colors. As many shades must be used as there are symbols in the pattern. When the background is imprinted as is usually the case, it may be filled in with any desired color, or left according to the material.

HALF CROSS STITCH. In Fig. 4 we have an example of half cross stitch. The stitches of one row are placed in the same direction over a long stitch of the wool or silk, which helps to cover the canvas but does not show on the right side. The manner of first laying the filling stitch and then working over it, is very clearly shown in the illustration.

By working row after row in the same direction the stitch is useful and rapid for filling in plain backgrounds, in sofa pillows, chair backs and rugs. But by chang-

Fig. 4.   Half Cross Stitch

ing the direction of the stitches variety is given to a monotone that produces the effect of colors and is often very effective. The manner of working a design in this method is quite clearly shown. In this example the canvas is to be left plain for a background. It may also be pulled away, leaving a design on the fabric on which the canvas was basted. Where the whole surface is to be covered in this stitch a scroll or key design is most suitable, worked in one direction while the background is filled in the opposite way.

GOBELIN STITCH. May be defined as a flat one that covers the canvas in parallel rows. One thread only may be covered by one stitch, or a dozen, or any number. Figs. 5 and 6 illustrate it fully. In Fig. 7 an example of slanting Gobelin is seen. Here three stitches are needed to fill each square. In Fig. 5 part of the canvas has been removed, showing the completed design upon the fabric. Gobelin stitch is sometimes called flat stitch.

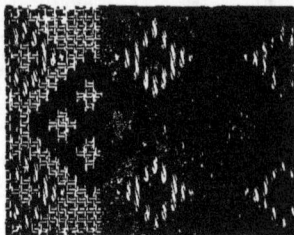

Fig. 5.   Gobelin Stitch

Any design that may be worked in common cross stitch, may also be carried out in slanting Gobelin. A pleasing method of working slanting Gobelin is to employ both silk and wool of the same shade in filling one square. The longer stitch may be of wool and the two shorter ones of silk.

RENAISSANCE STITCH. A very pretty and at the same time rapid stitch for covering large surfaces, such as sofa pillows or chair backs, is known as Renaissance stitch, and an exceedingly explicit illustration is given in Fig. 8. Two rows of the canvas are covered at the same time, and the method of doing so is shown with two colors, holding the needle in the two positions that give character to the stitch. Renaissance stitch is also suitable in working conventional designs on the large rug canvas.

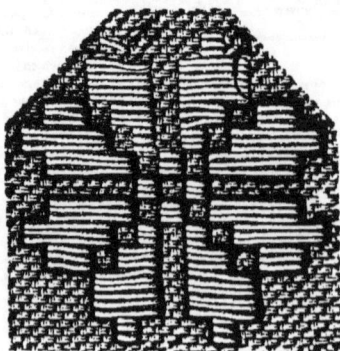

Fig. 6.   Gobelin Stitch

HOLBEIN STITCH. Another variety of cross stitch is popularly known as Holbein stitch, and looks very much like simple stitching. It is executed by taking up, and leaving alternate threads of the canvas or fabric. It is customary to continue it with simple cross stitch as in Figs. 9 and 10. Here the body of the each initial, is formed of the cross stitch, and the Holbein is used for a border and ornament. A detail of Fig. 9 is shown in Fig. 11. The

very rich alphabets, Figs. 55, 59, and 60 are worked in this manner. Also the ornamental borders, Figs. 62, 79, and 105, and the figures for ecclesiastical embroidery, Figs. 115, 116 and 123. Several other designs may be varied by adding outlines or finish in this stitch.

SQUARE STITCH. Figs. 12, 13, 14. This variety suggests Holbein at first glance. Two details are given showing the positions of each single stitch that forms the square. The dotted lines represent the crossing of the threads at the back. Many designs for single cross stitch may be executed in this manner, notably the borders, Figs. 2 and 15, and many figures and alphabets.

LEVIATHAN STITCH. Allied to simple cross stitch is the more complicated star or Leviathan stitch. Here a square of three meshes of the canvas is required to form one stitch or star. The method of placing the needle is shown in Fig. 16, while Fig. 15 exhibits an initial in the same stitch, on some square woven fabric. Any design for simple cross stitch may be doubled in size by working in star stitch. It is exceedingly effective when covering large surfaces, such as rugs, and borders on afghans or portières. Simple initials become ornamental in this stitch. Another application is as a filling stitch in the Russian work now so popular. The leaf or petal of a flower is covered with the stitch in such an open manner that the linen foundation shows through the interstices. The edge may be formed of Holbein or outline stitches. When an art linen, or any one of regular and rather coarse thread is used no canvas will be needed.

ROSETTE STITCH. Fig. 17, is a variety of the same stitch on the plan of four instead of three. It is commonly called star or rosette stitch. Fig. 18 shows the manner of working in which one half of a row is first worked and then the canvas is turned and the second half is worked back to the starting point. Fig. 17 shows a letter on linen in the same stitch. Very fine scrim may be basted in the center of a cambric handkerchief

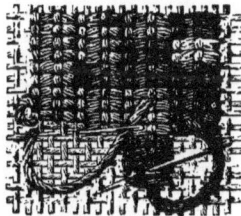

Fig. 8. Renaissance Stitch

and an initial worked in this stitch in fine embroidery cotton, with delicate and pleasing effect. Of course the scrim is pulled away, thread by thread when the work is done. This stitch has the same range and application as the one described above.

CHAIN STITCH. The adaptation of the familiar old chain stitch and cross stitch embroidery may be novel to many. The

regular spacing of the canvas serves as a guide to keep the work true and aids in copying the design. The stitch serves admirably for filling in grounds as shown in Fig. 19, and in the use it is often called "bird's-eye" stitch. In Fig. 20 we have a very handsome design in chain stitch over canvas on a plush or velvet foundation. When the canvas is pulled away a peculiar lace-like pattern appears. The effect is rich, delicate and novel. Much of the oriental embroidery is done in this stitch upon a coarse linen, that does not require canvas. The stitches are laid with the regularity of machine work, and are usually seen in masses. Fig. 21 gives a variety of this same stitch in a slanting position, which is rich and effective for

Fig. 9. Holbein and Cross Stitches

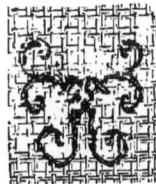

backgrounds. Many of the ecclesiastical designs on page 31 may be pleasingly varied by using this stitch, especially Figs. 120, 122 and 124.

PLUSH STITCH. An old-fashioned filling stitch is shown in Fig. 22 and is called "plush" stitch. The canvas is first covered with twin rows of common cross stitch leaving one row of canvas exposed between each set. As this part of the work does not show, any odds and ends of worsted may be used up in the process. But care must be taken that all is of the same size or an uneven surface will result. The wool for the outer surface is worked over the twin rows of cross stitch, pulling the needle through the holes in the canvas, first on one side and then

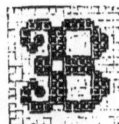

over the other, always taking up a thread which brings the point of the needle just back of a preceding stitch. The manner of weaving the threads back and forth In lattice fashion will be readily seen on referring to the illustration. Several rows of the lattice stitch are worked, one over the other, and the work is cut through the middle with a sharp knife when completed and evenly clipped, thus producing a ridged effect. When finished

Fig. 10. Holbein and Cross Stitches

a raised plush effect is produced that is exceedingly beautiful, when rich colors are employed.

KNOT STITCH. The method of forming a raised stitch that is familiarly known as a "French knot" and its application to canvas work is explicitly shown in Fig. 24. Here the wool is loosely wound around the needle twice, and the point is then inserted in the mesh of the canvas as in simple cross stitch; the direction in which the needle moves is shown by the dotted line, and arrow point. The thread may be wound around the needle as many times as desired. This stitch is useful for working rugs, cushions and similiar articles. In Fig. 23 we have an example of a border worked in this manner, in two colors, which might be applied to a frieze of a portière as well as to a rug. The cut illustrates the effect upon a fabric, but the stitch may be used to carry out almost any cross stitch

Fig. 11. Detail of Holbein Stitch

design on canvas with a background of the common stitch. A border in arrasene on a velvet or plush background would be exceedingly rich in this stitch.

SMYRNA STITCH. This stitch is used for rugs almost exclusively and is usually worked with double thread doubled so that

Fig. 7. Slanting Gobelin Stitch

four strands are employed, which are cut on the surface leaving a pile, somewhat similar to the well known machine rugs, hence the name. This effect may be reached in more than one way, and examples of the three principal methods are illustrated. Fig. 25 shows the way of first inserting the needle into the canvas so that the end of the thread shall be on the right side; while in Fig. 26 we have the easiest way of looping and tying the threads. Fig. 27 shows the same with some of the threads still uncut, and also the manner of working a design in different colors, and of forming a fringe or tassel border. Another method of making this stitch is shown in Figs. 28 and 29. The needle is first inserted as in Fig. 25, and drawn until the threads extend on the right side for about one-half or three-eighths of an inch, and is then carried to the mesh that is next higher and brought out of the one in which it was first inserted. The method of dividing the strand around the tuft which lies or holds it in place can be best understood by studying the illustration. A continuation of the process is shown in Fig. 29, and the starting point of a new stitch is shown by the arrow-head. When the rug is covered, the stitches must be carefully and evenly trimmed and then combed out.

The method of carrying out a cross stitch design with pieces

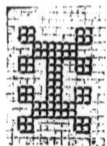

Fig. 12. Square Stitch

Fig. 13. Detail of Square Stitch

Fig. 14. Square Stitch Completed

of cloth is shown in Fig. 30. The pieces may be cut in rounds, oblongs or stars, and are sewed to the rug canvas by a strong carpet thread, which however, does not show when the pieces are drawn tightly into place. A hit-or-miss center with a border in colors, black foundation, using a simple scroll or key design is the most satisfactory manner of using this method. There is no objection to copying the more intricate designs of a woven rug, but that requires more skill. The pieces may be sewed on coarse burlaps coffee sacking, that is woven in a square mesh instead of on the regular rug canvas. The latter is preferable for elaborate designs.

We have thus far considered the various stitches and their modifications, that may be used in cross stitch embroidery. We will now extend their application to various fabrics, striving to cover the whole field. First we will consider such materials woven with a square thread that do not need the aid of the ordinary Berlin or Penelope canvas in counting the stitches and keeping them true.

Fig. 15. Leviathan Stitch

One of the first that suggests this method is the familiar Java canvas. This may be used as the entire article for sideboard, bureau and stand covers, and for cushions and tidies, or it may be cut in strips and employed as a decoration to some other material. The background is left open. Congress canvas or scrim is also familiar to all who have embroidered in this stitch. It is a favorite with the Germans. It is used without filling in the background, and to form the whole article or is employed in bands and applied either as inserting or upon some other material. In the present fad for decorating garments with narrow bands of cross stitch embroidery, bands of this scrim are extremely popular. Fig. 31 shows one figure that powdered a table cover of scrim, worked in three colors. The edge was finished with a buttonhole scallop. Many of the borders illustrated in this manner, are beautiful for this purpose, especially Figs. 84, 85 and a half of 90 or 96. For the same use Figs. 103 and 161 may also be used. Scrim may be basted upon ribbon or any fabric, as will be described later, and the threads pulled away after the work is completed. For simple border patterns upon congress canvas or scrim of ordinary mesh, rope

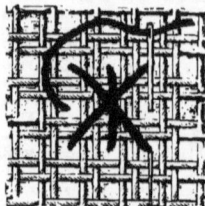

Fig. 16. Detail of Leviathan Stitch

Fig. 17. Rosette Stitch

Fig. 18. Detail of Rosette Stitch

silk is at once one of the richest and most beautiful threads to employ. There are also several silk or French canvases.

Besides these canvases there is a close woven one of a beautiful cream color, to be worked with silk; and an imitation of the old Gobelin, but in cotton, and bearing the same name. Art linen comes in buff, cream and white, and is very firm material. There are several other linens, from the coarse, heavy butcher's, to the sheer French linen, and as many damasks. Then comes duck in cream and white, and in fancy weave, and twill and drill, either dark blue or white, and suggestive of Russian work. All these are suitable for cross stitch embroidery and may be embroidered with silk, but usually require cotton or linen thread.

Fabrics that rank next to the canvases in ease of working are ginghams, cheviots and checked goods. The example (Fig. 32) shows a pretty design in cross stitch upon gingham, which is suitable for trimming a dress or an apron

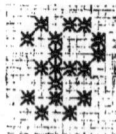

Fig. 19. Chain Stitch

This is worked in white linen. Cheviot is usually worked in crewel or silk, and applied as a border on wool gowns.

A material that may be used without canvas with success for bordering gowns or top garments is bunting. A cream ground is a general favorite, as it shows off rich colors admirably. The

Fig. 20. Chain Stitch

regular weave renders the material almost as easy to work on as canvas. It should be worked in a frame. The borders shown in Figs. 127 and 160, as well as the ones mentioned above are adapted to this style. The bands should be fastened to the garment with fancy feather stitches.

Cheese cloth or batiste is also used as a foundation for the popular borders in cross stitch. The design is worked upon strips which are used for trimming children's aprons and dresses.

Fig. 21. Slanting Chain Stitch

Fig. 22. Plush Stitch

In Fig. 33 we have an example of a band worked in two colors upon white braid, or a strip of bunting or similar material, and the method of applying it to the garment with fancy stitches. Rope or embroidery silk is generally used upon woolen or silk foundations. For dainty work, tarletan may be laid over satin

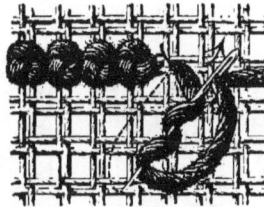

or ribbon, and worked with floselle and gold thread. An exquisite center for a toilet cushion might be fashioned in this way, using such square designs as given in Figs. 88 or 150, or the quaint griffin in Fig. 145.

Or by employing ribbon in harmonizing colors working in rich yet simple designs and joining the strips together, lovely fan bags might be made. Such embroidered strips might also be used in decoration of dinner dresses, and tea gowns in Russian style.

A method of varying these borders is to work the background only leav-

Fig. 23. Knot Stitch

ing the canvas or bunting to show in the design. This is seen in Fig. 34. In working on linen also, or upon towels, this method yields a pleasing variety. Care will be needed in selecting patterns that give connected masses as these only are successful. Those borders shown in Figs. 57, 62 and 75, and several others, would give satisfaction in this style. Much of the true Russian embroidery is worked in this manner. It is sometimes desirable to decorate straw goods with embroidery. Fig. 35 shows a pattern worked upon a basket in three colors in star and simple cross stitches.

Common net as well as the newer fish-net, may also be decorated with cross stitch, and used for borders or hangings. A dainty toilet table may be fashioned in this manner. The hang-

Fig. 24. Detail of Knot Stitch

ings are powdered with set designs like the cross in the accompanying cut (Fig. 36) or the clover leaf taken from border Fig. 70, or any simple flower. The lace may be worked with the border Fig. 101, in one color of course, buttonholing the edge and cutting out the scallop.

GUIPURE D'ART. Filet and grenadine are materials that serve as foundations for some exceedingly dainty embroidery. The

Fig. 25. Detail of Smyrna Stitch

Fig. 26. Detail of Smyrna Stitch

stitch belongs more properly to the darning or lace order, than to cross stitch embroidery. But as articles so decorated require similar designs, it is illustrated here. In Fig. 37 we see the method of weaving the thread back and forth, and in Fig. 38

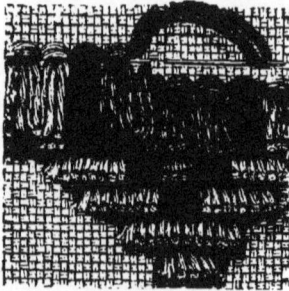

Fig. 27. Smyrna Stitch

the completed design. It is used for the decoration of dresses, crowns of hats and for many of the purposes of net.

BASTING CANVAS UPON FABRICS. The first step and a very important one in attempting to embroider in cross stitch upon plush, silk, broadcloth, or any material that is not woven in

Fig. 28. Detail of Smyrna Stitch

Fig. 29. Detail of Smyrna Stitch

prominent or regular threads, is the method of properly basting Penelope canvas, scrim or a similar material upon the chosen surface. This is neither so easy, nor so simple as appears.

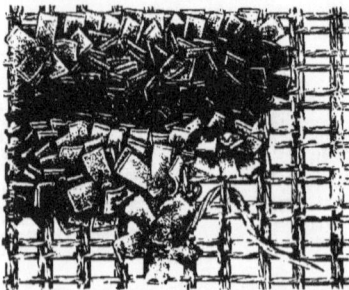

Fig. 30. Method of Working Rug

First the canvas must be placed upon the fabric squarely thread by thread, or when it is afterwards removed the design will be all

askew. Then it must be basted securely so that it will not slip in working. Care is especially necessary in placing it on plush or velvet. Fig. 39 shows the proper manner of basting. Cross stitches are used at intervals to prevent slipping. In Fig. 40 we have a strip of canvas basted upon twill, for one of those Russian aprons shown in Fig. 2 and partially worked. In working, care must be taken to put the needle through both the canvas and the cloth, otherwise when the canvas is removed the stitch will be loose, and that portion of the work will have to be done over again. When a frame is used, and the stitches are taken back and forth, there is little danger

Fig. 31. Cross Stitch on Scrim

of this vexing mistake happening. Another fault to be guarded against is bringing the needle through any of the threads of the canvas itself; if this should happen it will be difficult to draw away the thread when the work is finished without disarranging some of the cross stitches. In this case it is well to draw the thread and cut it where it is caught down by the stitch, than to finish removing it by drawing it in the opposite direction. It is best to use the common embroidery hoops whenever possible. This removes the possibility of warping the foundation; the

Fig. 32. Cross Stitch on Gingham

threads can also be drawn tighter, so that the stitch will not be too loose when the canvas is pulled. When the design is completed the canvas must be taken away with great care. The first thing to be done is to remove the tacking threads which hold the canvas down to the material; the canvas is then pulled away thread by thread so carefully as not in any way to disturb the set of the stitches. Some of the patterns are so complicated there is no opportunity of cutting away any part of the canvas to render the task of removing the threads less tedious, as may often be done in the case of less well-covered designs. It is as well to draw the shortest threads first, for it is difficult to take out the long ones along the whole length of the strip until this is done. The worker will find, too, that it does not answer to try

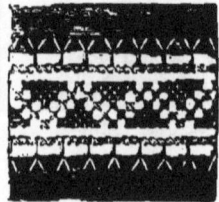

Fig. 33. Cross Stitch on Braid.

to draw out more than one, or at most, two threads at a time until nearly all in one direction have been removed. She will also see that too much vigor is apt to disturb the embroidery, which has to be worked rather tightly so that it shall not set too loosely against the material when the canvas threads have been drawn away. When the canvas is all removed the work must be pressed with a moderately warm flat iron. Workers must notice that I advisedly use the word "pressed" instead of "ironed," for

Fig. 34. Cross Stitch used as a Background

a too vigorous ironing, without due regard to the general "set" of the work is likely as not to force it out of shape, and thereby to render it crooked. A skillful hand will manage the iron so that it does away with existing imperfections instead of creating fresh ones. If the iron is too hot the colors of embroidery worked with colored cotton will be apt to be affected by it, and the yellow more especially, though the makers guarantee that it will wash thoroughly well, is likely to deepen in tone under the influence of the heat. These general directions apply to ordinary fabrics that require the use of canvas. Plush and velvet are the exception. Here the stitch must always be taken perpendicularly and the canvas need not be removed in the case of separate figures. We have an example in Fig. 41. In this case, work the figure

Fig. 35. Cross Stitch on Straw

to the last row; then cut the canvas in such a way that the last stitches cover the ends of the threads. This saves the trouble of pulling out the threads. Very elegant effects may be produced in rope silk upon plush. Of course this work must never be pressed as that destroys the elasticity of the pile and the beauty of the plush. Therefore great care must be taken not to draw or warp the fabric while working the design. A variation of this method is to trace the pattern upon the canvas after it is basted upon the cloth, by means of long horizontal stitches. Afterwards the whole is worked over in half cross stitch, and every feature of the design is well shown up. See Fig. 42. The best imported designs are usually worked in this manner. Split zephyr or crewel, in the proper colors, are used for the foundation, and are worked over with

heavy silk or wool. This does away with the necessity of counting every stitch in the second working. Gobelin stitch may also be used, and on a heavy fabric like broadcloth is particularly

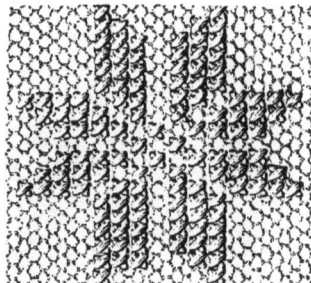

Fig. 36. Cross Stitch on Net

rich. The canvas threads may be pulled out or cut away as described above.

CROSS STITCH IN CROCHET. A simple way of working cross stitch patterns is in crochet. In this method the open work is formed in squares (on a chain foundation) by working 1 t c in a ch, ch 2, miss 2 ch, 1 t c in next ch, etc. The solid squares form-

Fig. 37. Detail of Darned Net

Fig. 38. Darned Net

ing the design are made by having 2 t c between, instead of the 2 ch. Any cross stitch pattern may thus be followed. A narrow border design with scallop added, makes a pretty lace, the border alone forming the insertion. Wide borders set together with ribbon or embroidered bands of linen may be used for bed-spreads, table covers, curtains, etc; while squares and other large designs may be used for chair-backs, cushion covers, etc. One has only to remember that the open squares are formed of 2 t c separated by 2 ch, and the solid squares of 4 t c, placing the two t c between, instead of the 2 ch. Of course these proportions can be varied at pleasure. The squares shown in Figs. 149 and 150 are suitable for bedspreads in crochet. We have an exam-

Fig. 39. Canvas basted upon fabric ready for working

ple of using cross stitch patterns in crochet work in Fig. 43, which is simply a detail of a larger design. The manner of working it clearly shown. A very handsome design is given in the next cut (Fig. 44) which may be used for any of the purposes mentioned and also for insertion in a short curtain of scrim.

In these examples, but one color is used, and the foundation of the design is left in open work. But in the next two cuts (Figs.

tion may be the usual canvases and fabrics, or the pattern may be followed in knitting or crochet. In the latter case the beads are first strung; and then knit or crocheted in place In the same manner in which a design is worked out, and which has already been explained. The style is particularly suitable for silk and

Fig. 40. Strip of Russian Apron in process of working

45 and 46) we have an example of solid work in short crochet, and the method of working this design, requires three colors and the manner of carrying one thread at the back of the work too well know to need description. Fig. 46 gives a section of the completed strip, which is especially suitable for an afghan. It may also be carried out in tricot, as the squares formed by that stitch form a good guide for working cross stitch patterns.

Short crochet is a stitch much used in making half shawls of Germantown. The middle may be of white or a solid color, and the border in black and several shades of one color. The rich border, Fig. 93, might be crocheted in four shades of soft wood browns or any color upon a white ground, using eight rows of the lightest shade for the top, and graduating by eight rows each,

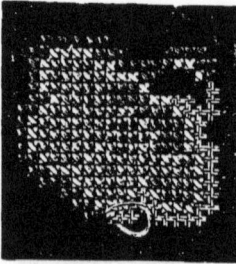

Fig. 41. Cross Stitch Applique

to the darkest at the bottom. A row or two of black may be crocheted above and below as a finish. Fig. 92 Is adapted to the same purpose.

KNITTING AND CROSS STITCH. In working a cross stitch design by knitting, common garter and purl stitches and two needles are used. The ground is formed by the garter stitch, and the design by purling one for every square in the pattern. Two colors may be used and the thread carried behind in the usual manner. We have a very perfect example in Fig. 47, which might be used for a bedspread, or any of the purposes mentioned for crocheting cross stitch patterns.

BEAD WORK. A pleasing way in which to enrich a pattern in cross stitch, is by the introduction of beads. One bead is used in the place of each square in the design. The founda-

bead purses and the like. Simple figures, initials or monograms (Figs. 48 and 49) may be worked with beads. Sometimes spangles and beads are mingled with cross stitch in the same decoration as seen in Fig. 50, which is a border on a drapery of scrim.

In examining the varied and beautiful designs that are printed in this book, many uses are discovered that have not yet been mentioned. The numerous alphabets are particularly handsome for marking blankets and towels. It is a fad worthy of becoming a custom, to copy the Germans in working appro-

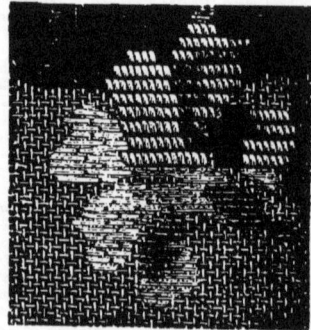

Fig. 42. Cross Stitch Applique

priate sentiments upon the serviettes and other pieces of table linen. For this purpose, such alphabets as 64 and 66 are suitable.

These mottoes are placed above or between rows of embroidery. A square that may be used for center piece, five o'clock tea cloth, and other purposes, is edged with German lace, above that a rich border in cross stitch running around the four sides, and inside of that a motto, one line on each side

"Rein Sie die Liebe,
Rein Sie der Mund,
Rein Sie der Trank,
Das Herz gesund."

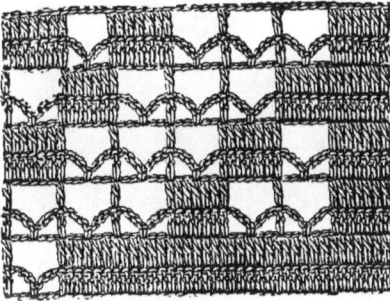

Fig. 43. Detail of Cross Stitch Design worked in Crochet

Which may be translated:
"Pure is the appetite,
Pure is the food,
Pure is the drink,
That God gives."

A serviette with a wide border, composed of several rows of harmonizing designs and drawn work, bears this motto on one end:

"Salz und Brod Gebe Gott."

And on the other the initials and date in Roman numerals.
Another motto in old German reads:

"An Got nit Verzag Glueck kombt Alle Tag."

A free translation is something like this:

"Never despair, luck comes every day."

The ornate capitals which are given in four very rich and different styles, are exceedingly effective in working initials upon broadcloth carriage robes or afghans, handbags or shawl cases. Many of the designs are appropriate for the Russian towels that now

Fig. 44. Cross Stitch Design worked in Crochet

grace every guest room with pretense of elegance.
These red and blue embroideries appear on all household and personal linen, especially on the large napkins or towels which are found alike in the very poorest Russian huts and in the royal palaces. Some of these are heirlooms in peasant families, and they are highly ornamental, because on fête days they are hung up as decorations. The embroidery of a true Russian napkin or towel will be in three divisions, sometimes separated by bands of colored linen

Fig. 45. Detail of Cross Stitch Design worked in Crochet

or cotton and lines of drawn work, of which the central or principal one is the design proper, and the narrower on either side are called the frieze or cornice. For these three

Fig. 46. Section of Cross Stitch Design worked in Crochet

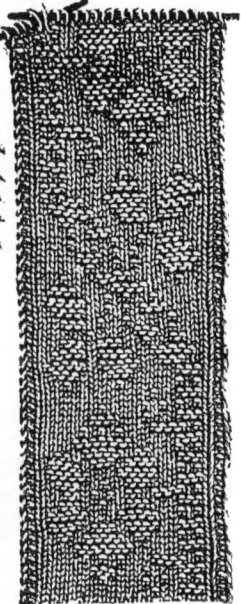

Fig. 47. Cross Stitch Design carried out in Knitting

divisions we select Figs. 67, 79 and 105, working them in the order named and in the conventional colors. For marking such a towel, letters will be chosen from alphabets, Figs. 55 or 60.
Fig. 1 at the beginning of the book illustrates these sham German towels, with loops for hanging on the wall.
The quaint patterns given in Figs. 70, 111, 112 or 156 may be

Fig. 48. Cross Stitch Design worked in Beads

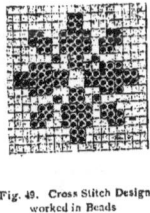

Fig. 49. Cross Stitch Design worked in Beads

taken for the motif, and any of the narrow bands added at pleasure. Figs. 106, 140 and 148 are also strongly German in style. Another of the many designs adapted to these very fashionable

12

sham towels is Fig. 104. Many others are scattered over the pages.

For a nursery blanket or a crawling rug, Fig. 97 may be chosen,

Fig. 50. Cross Stitch Design worked in silk, beads and spangles

the designs being worked on a border of burlap or appliquéd on cloth while Figs. 98 and 99 at once suggest chair backs and afghans.

For ecclesiastical embroidery a number of very rich designs are shown from Figs. 113 to 124. These are appropriate for stoles, chasubles, altar cloths, etc.

A very beautiful model is shown in Fig. 52.

The classic cross given in Fig. 125, is appropriate for the cover of a prayer book, altar frontal, lectern hanging or Bible marker. The size of the canvas must be varied for these different uses.

Fig. 51. Serviette

A very rich design for a super-frontal for the altar may be worked with Fig. 131, in several colors and gold thread. The proper symbols are employed to designate each color.

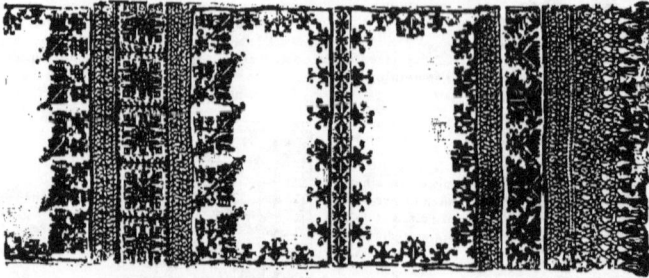

rug design is seen in Fig. 139. which is worked on the large rug canvas in Smyrna stitch. It may also be copied in knot stitch, or in knitting. The different colors are expressed by symbols. For a child's bib the humorous designs, Figs. 140 and 147 never fail to please the wearer.

Fig. 52. Altar Cloth

A magnificent piece of embroidery was a plush slumber roll worked with Fig. 154 in rope silk. The all-over designs given in Figs. 155, 157 and 158, and elsewhere are used for bags and for articles of burlap canvas.

An elaborate "runner" for the dinner table is given in Fig. 54, showing two-thirds of the whole. A similar choice and arrangement of designs might be used for a bedspread. A wide border should be worked from top to bottom through the middle of the piece of linen, and horizontal stripes on each side as in the cut. One large sheet of linen may be used for such a spread, or for convenience, it may be composed of separate strips joined. Don't forget to work initials and date of year. Such a spread would require leisure and perseverance, but would be an heirloom when finished. Fringe may edge the sides or heavy German lace.

There is scarcely a pattern in this book that may not be applied to a dozen different uses or materials, or worked in many different stitches. And here lies one great charm of

Fig. 54. Table Runner

The design at Fig. 135 is suitable for a chair seat, although the vines may be used for many other purposes. For a sofa pillow nothing could be more beautiful than Fig. 137. A quarter of a

cross-stitch embroidery, the latitude enjoyed by the worker.

And in the frequent revivals of this stitch we have certainly an example of the "survival of the fittest."

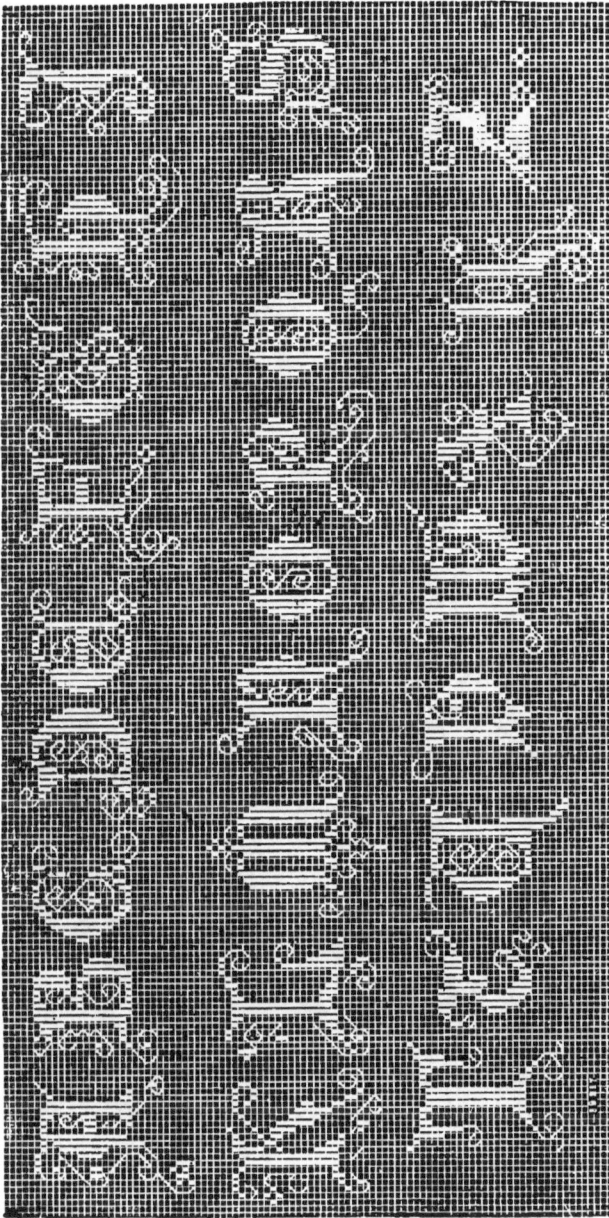

Fig. 55

Fig. 56

Fig. 57

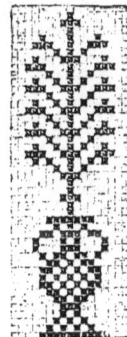

Fig. 58

Fig. 59

Fig. 60

Fig. 61

Fig. 62

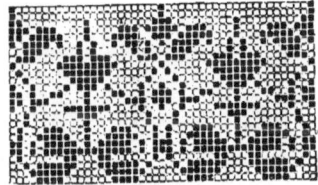

Fig. 63

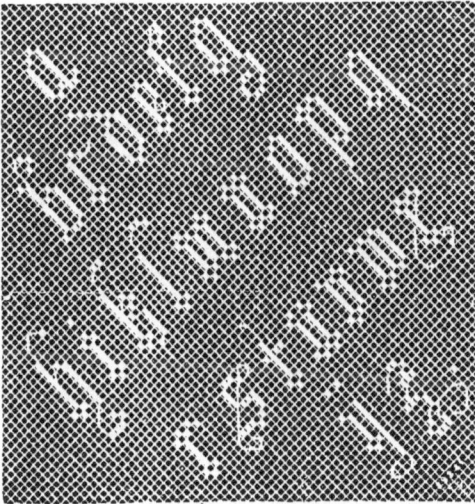

Fig. 64

Fig. 65

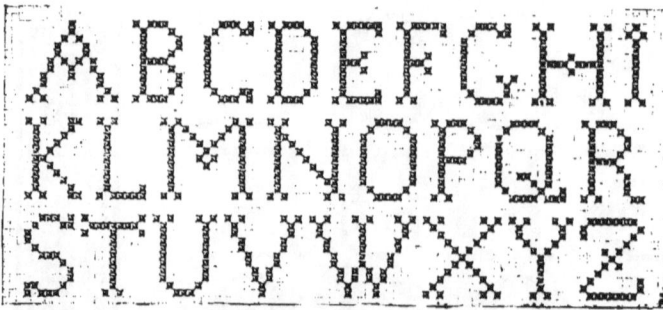

Fig. 66

Fig. 67

Fig. 68

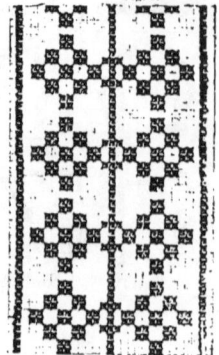

Fig. 69

Fig. 70

Fig. 71

Fig. 72

Fig. 73

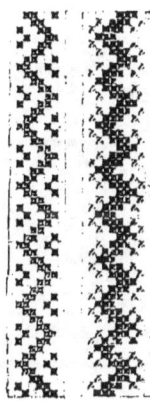

Fig. 74   Fig. 75

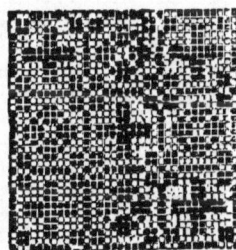

Fig. 77

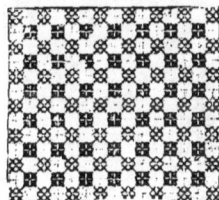

Fig. 79

Fig. 80.

Fig. 81

Fig. 82

Fig. 83

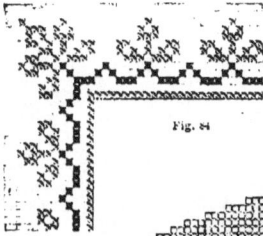

Fig. 84

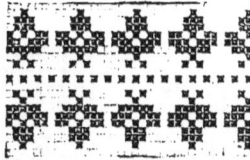

Fig. 85

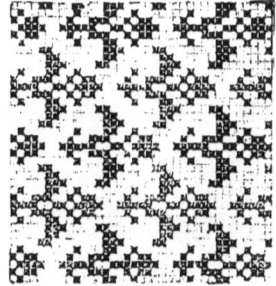

Fig. 86

Fig. 87.

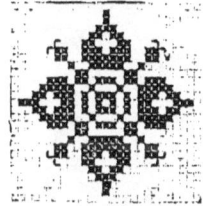

Fig. 88

Fig. 91

Fig. 89

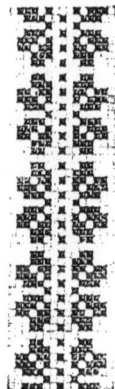

Fig. 90

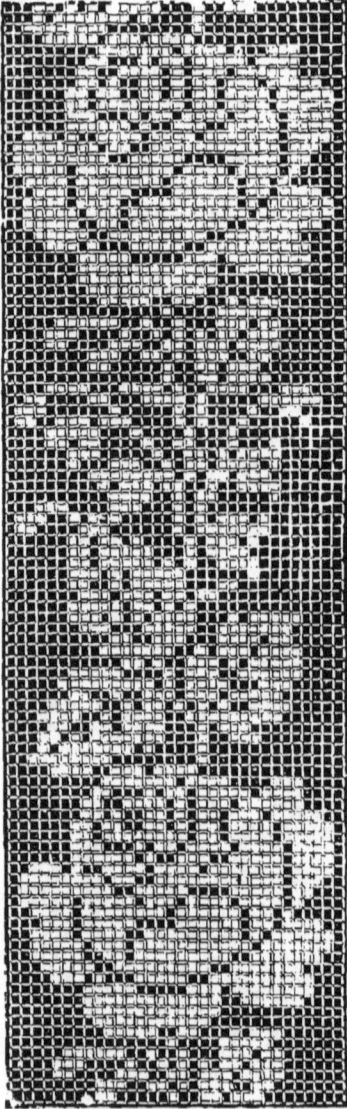

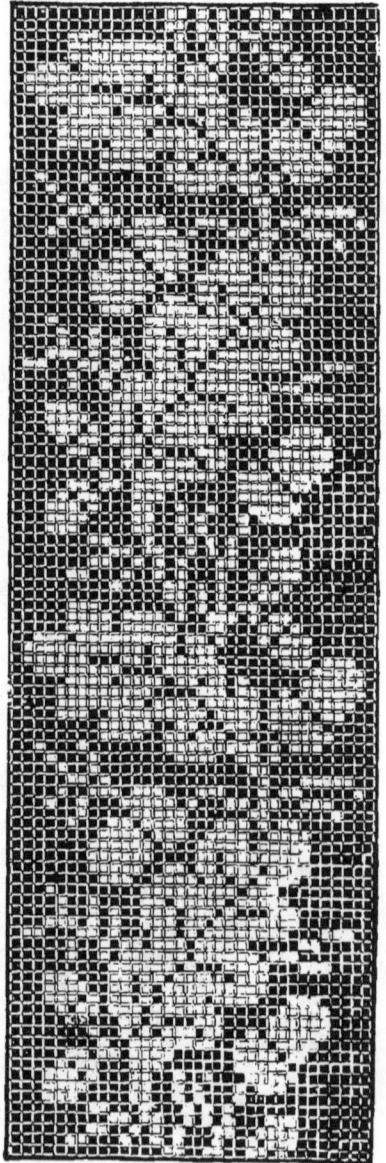

Fig. 92

Fig. 93

Fig. 94

Fig. 95

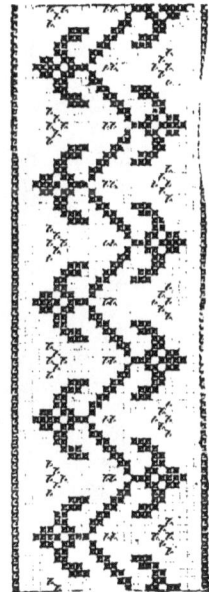

Fig. 96

□ Dark Red.   ◆ Medium Red.   ▣ Light Red.   ■ Dark Brown.   ▨ Light Brown.   ▩ Dark Yellow.   ✕ Light Yellow.   ▣ Blue
□ Green.   ● Dark Gray.   ▥ Light Gray.

Fig. 97

■ Dark Brown. ⊡ Bronze Brown. ▯ Dark Gray. □ Light Gray. ⊠ 1st, (Darkest), ⊠ 3d, ⊡ 3d, ⊡ 3d, (Darkest) Blue. ⊠ 1st, (Darkest), ⊠ 2d, ⊡ 3d, (Lightest) Red. ⊡ Olive Green. | White, (Ground).

Fig. 28

■ 1st, (Darkest), ⊠ 2d, ▯ 3d, ⊡ 4th, □ 5th, (Lightest) Red. ■ 1st, (Darkest), ⊠ 2d, ⊡ 3d, ⊡ 4th, (Lightest) Green. □ 1st, (Darkest), ⊡ 2d, ⊡ 3d, (Lightest) Blue. | Ground.

Fig. 29

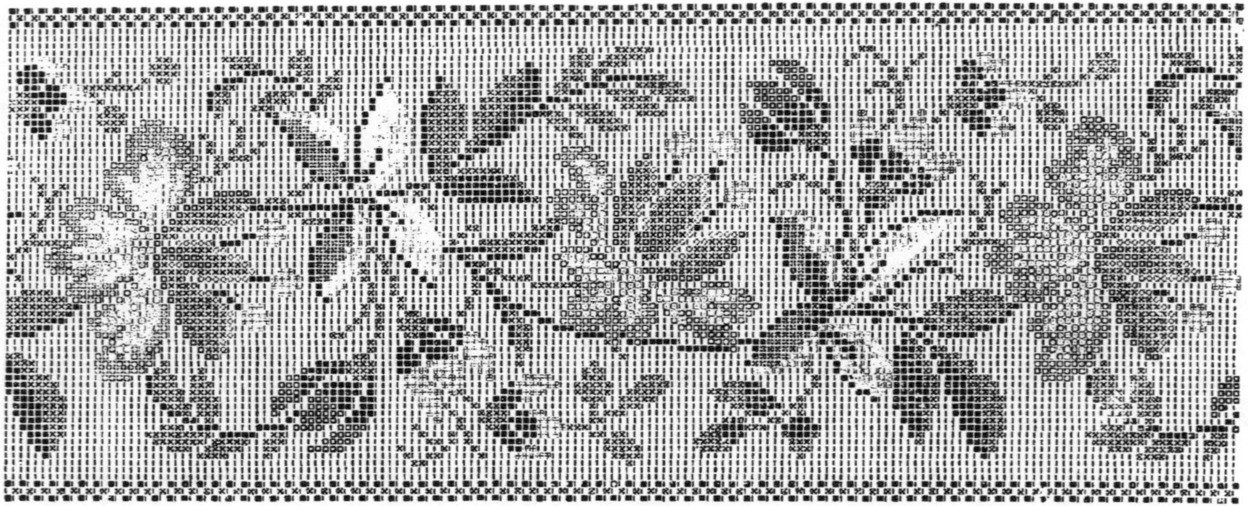

1st, (Darkest). 2d. 3d. (Lightest) Olive Green. 1st, (Darkest), 2d, 3d, (Lightest) Wood Brown. Dark Red. Light Red. Dark Blue. Light Blue. | Ground.

Fig. 100

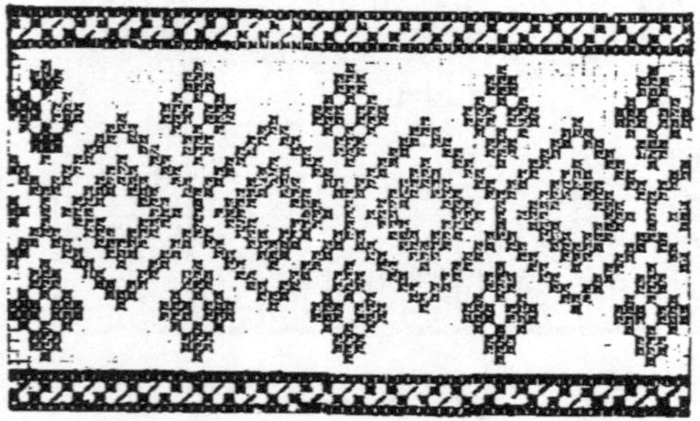

Fig. 102

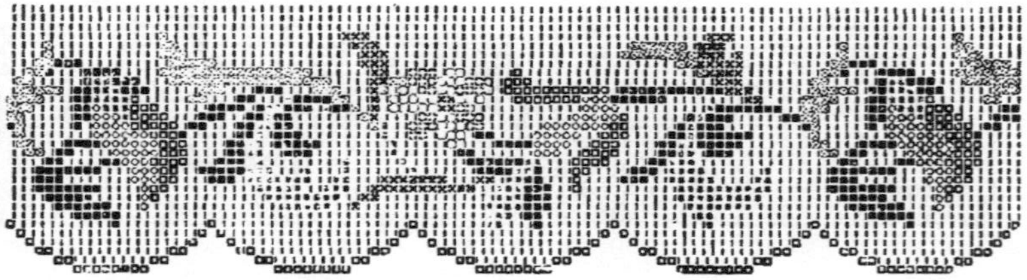

Fig. 101

Dark Olive. Light Olive. Dark Green. Light Green. Dark Blue. Light Pue. Dark Red. Light Red.

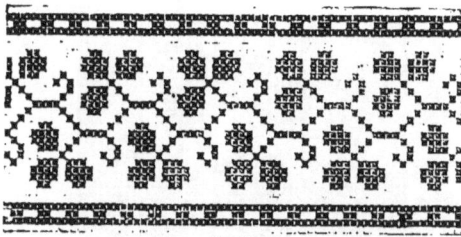

Fig. 103

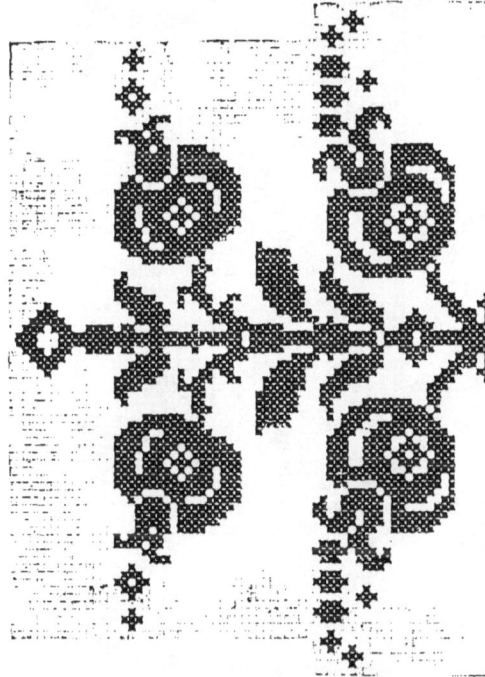

Fig. 104

Fig. 105

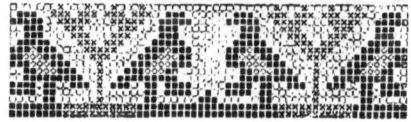

Fig. 106

■ Black. ☐ Corn Color. ▨ Peacock Blue. ◎ Bronze. ✕ Maroon. ✕ Olive.
Fig. 167

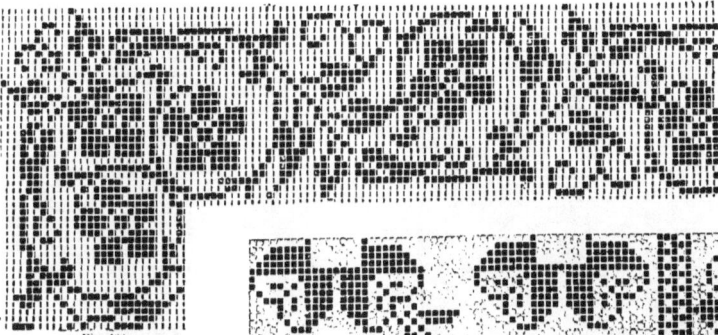

Fig. 108.

Fig. 109

Fig. 110

Fig. 111

Fig. 112

Fig. 113

■ Dark Brown. □ Bronze Brown. ▣ Gray. ⏐ Ground.
Fig. 117

Fig. 120

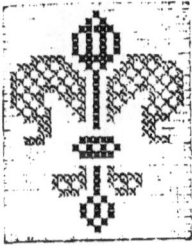

Fig. 114

Fig. 121

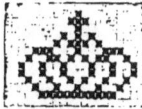

Fig. 118

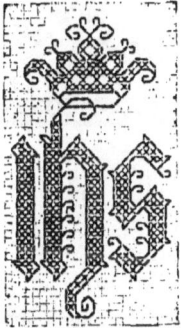

Fig. 115

Fig. 119

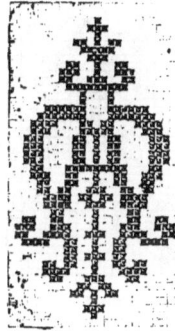

Fig. 122

Fig. 123

Fig. 116

Fig. 124

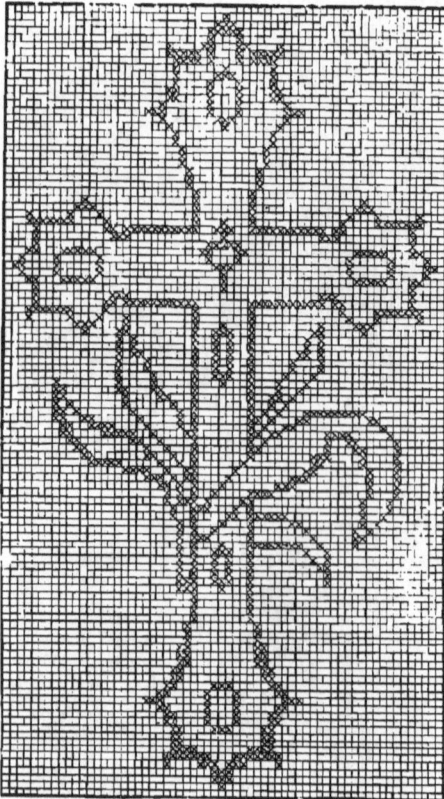

Fig. 125

Brown, ◫ 1st, [Darkest], ▨ 2d, ▯ 3d,
[Lightest] Green ◫ 1st, [Darkest] ◫ 2d, ◪ 3d,
[Lightest] Red, ◫ 1st, [Darkest] ▨ 3d, ◪ 3d.
[Lightest] Gray. ▯ Yellow, | Ground

Fig. 126

Fig. 127

Fig. 130

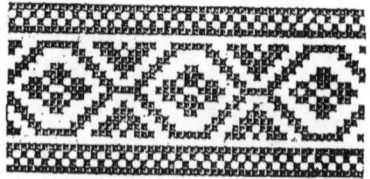

■ 1st, [Darkest], ◫ 2d, ▯ 3d, ◻ 4th, [Lightest]
Blue, ◫ 1st, [Darkest], ◫ 2d, ◪ 3d, [Lightest]
Brown, ◫ 1st, [Darkest], ◫ 2d, ◫ 3d, [Light-
est] Olive Green. ✕ Bronze Brown.

Fig. 128

■ 1st, [Darkest], ◫ 2d, ═ 3d, ◻ 4th, [Lightest]
Lilac. ◫ Dark Yellow, ◪ Light Yellow.
◫ 1st, [Darkest], ◫ 2d, ◪ 3d, [Lightest]
Olive Green. ✕ Bronze Brown.

Fig 129

■ Brown, ◫ 1st, [Darkest], ✕ 2d, ▯ 3d, [Lightest]
Green, ◫ 1st, [Darkest] ◫ 2d, ◪ 3d, [Lightest]
Gray, ◫ 1st, [Darkest], | 2d, ▨ 3d, [Lightest]
Blue, ◻ Yellow, | Ground.

Fig. 130

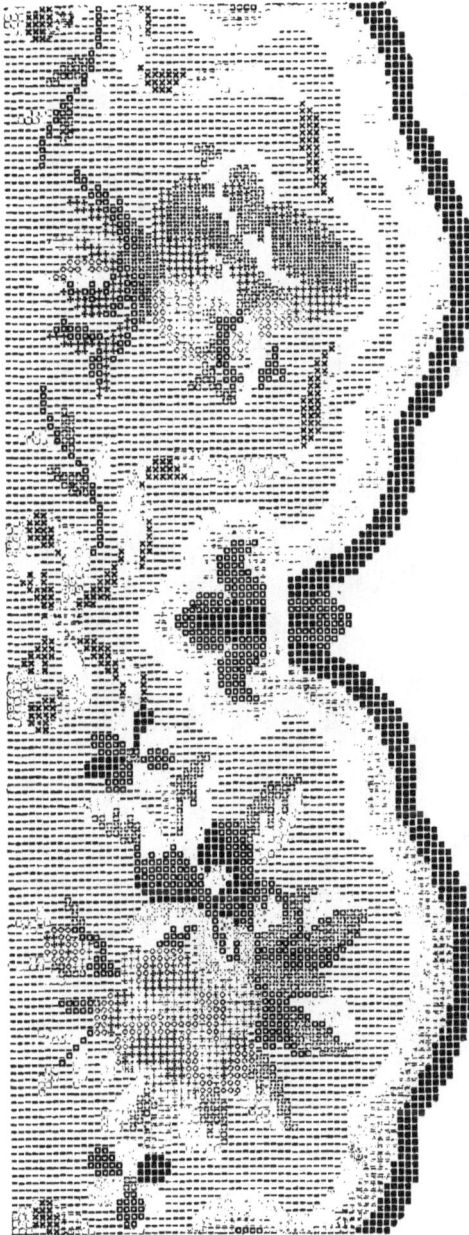

Fig. 131

Fig. 132

Fig. 133

Fig. 134

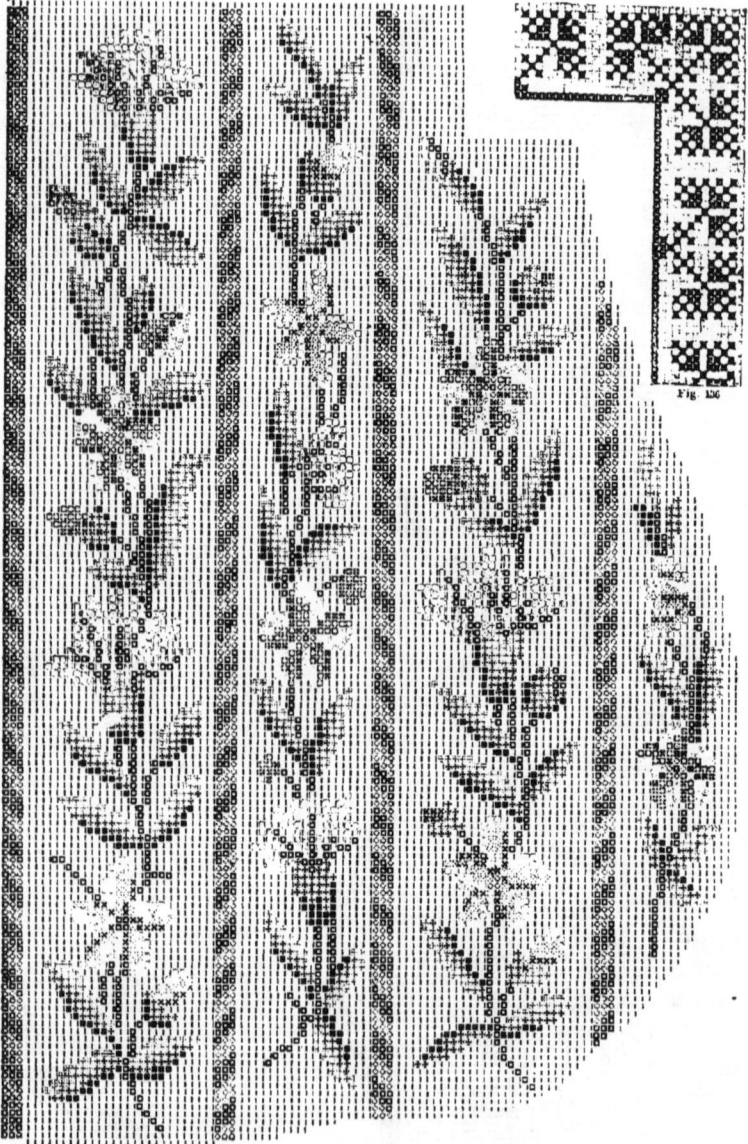

Fig. 136

⬛ 1st (Darkest), ⬛ 2d, ⬚ 3d, (Lightest) Lilac. ⬛ 1st, (Darkest), ⬛ 2d, ⬛ 3d, (Lightest) Blue-Green. ⬚ Old Gold. ⬚ Bronze Brown. ⬚ 1st. (Darkest). ⬚ 2d, ⬚ 3d, (Lightest) Blue. ⬛ 1st, (Darkest), ⬚ 2d, ⬚ 3d, (Lightest) Red. ⎹ Ground

Fig. 135

35

Fig. 138

Black. ⊡ Yellow. ⊞ Green. ⊟ Red. ⊟ Blue. ⊠ White.

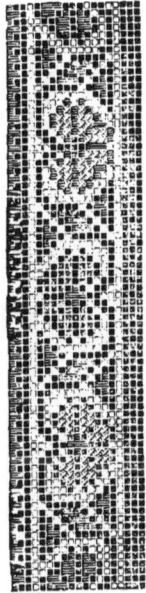

▤ 1st. (Darkest). ▯ 3d. ▨ 4d. (Lightest) Olive Green. ▦ 1st. (Darkest). ▨ 2d. ▨ 3d. (Lightest) Wood Brown. ▨ Dark Red. ▨ Light Red. ▨ Dark Blue. ▨ Light Blue. | Ground.

Fig. 137

▨ Dark Brown. ◉ Dark Red. ▨ Light Red. ▨ Dark Blue. ▨ Light Blue. ✕ Dark Olive. ▨ Light Olive. ▨ 1st. (Darkest). ▨ 2d. | 3d. (Lightest) Gray.

Fig. 139

Fig. 140

36

ist, (Darkest), ☐ 2d, ▐ 3d, ☐ 4th, (Lightest) Green. ▐ ist, (Darkest), ▐ 2d, ☐ 3d, ☐ 4th, (Lightest) Brown. ☒ ist, (Darkest), ☐ 2d, ☒ 3d, │ 4th, (Lightest) Gray.

Fig. 141

Fig. 142

Fig. 143

■ ist, (Darkest), ☐ 2d, ▐ 3d, (Lightest) Red. ☒ Pink. ☒ Dark Brown. ▩ Light Brown. ☐ Gray. │ Yellow. ▐ ist, (Darkest), ☐ 2d, ▩ 3d, (Lightest) Blue. ☐ Olive Green.

Fig. 144

Fig. 145

Fig. 146

Fig. 147

Fig. 148

Fig. 149

Fig. 150

Fig. 152

■ 1st, (Darkest), ☐ 2d, ☐ 3d, ☐ 4th, (Lightest) Green. ☒ 1st,
(Darkest), ☐ 2d, ☒ 3d, ☐ 4th, (Lightest) Red. ✕ 1st, (Darkest),
■ 2d, ☐ 3d, ☐ 4th, | 5th, (Lightest) Brown.

Fig. 151

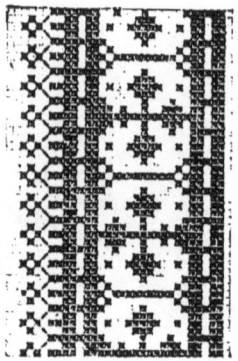

Fig. 153

■ 1st, (Darkest) ✦ 2d, ▨ 3d, ☐ 4th, ✕ 5th, (Lightest) Red. ☐ 1st, (Darkest), ✗ 2d, ▨ 3d, ☑ 4th, (Lightest)
Green. ▤ Dark Yellow. | Light Yellow.

Fig. 154

Fig. 155

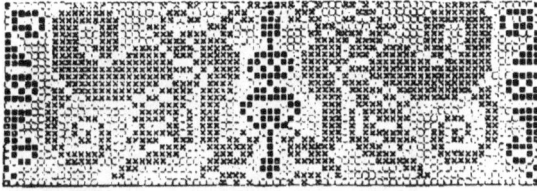

Fig. 156

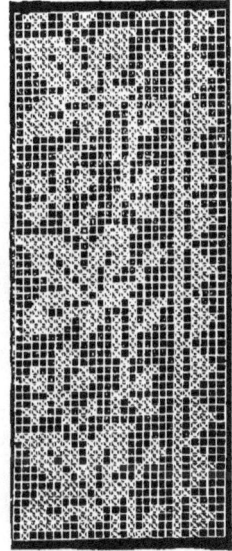

Fig. 157

Fig. 159

Fig. 158

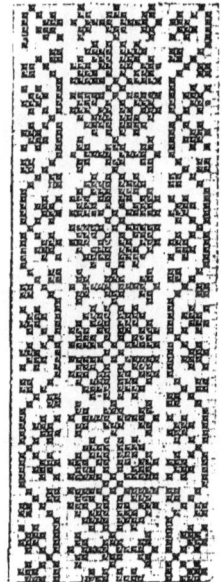

Fig. 160

Fig. 161

■ 1st, (Darkest), ⊠ 2d, ⊞ 3d, (Lightest) Reddish-Brown. ■ 1st, (Darkest)
⊠ 2d, ║ 3d, ⊡ 4th, │ 5th, (Lightest) Fawn.

Fig. 162

Fig. 163

■ 1st, (Darkest), ✕ 2d, ⊠ 3d, (Lightest) Brown. ⊞ 1st, (Darkest), ⊞ 2d,
□ 3d, (Lightest) Blue. ⊠ 1st, (Darkest), ⊞ 2d, ⊠ 3d, ⊠ 4th,
(Lightest) Green. ⊠ 1st, (Darkest), ⊞ 2d, ⊠ 3d,
(Lightest) Red. │ Gray.

Fig. 164

Fig. 165

TABLE COVER IN CROSS-STITCH WITH DETAILS OF PATTERN

Fig. 166. CHAIR-BACK

■ Dark Green. ⊓ Dark Red. ⊑ Pink.
DETAIL of Fig. 166

Fig. 167. MANTEL DRAPERY

DETAIL OF Fig. 167   ■ Dark Blue. ✖ Green. ▨ Light Brown. ● Dark Brown. ▨ Terra-cotta. ■ Middle Blue.
✕ Lightest Blue. ▨ Yellow. ▫ Pink.

43

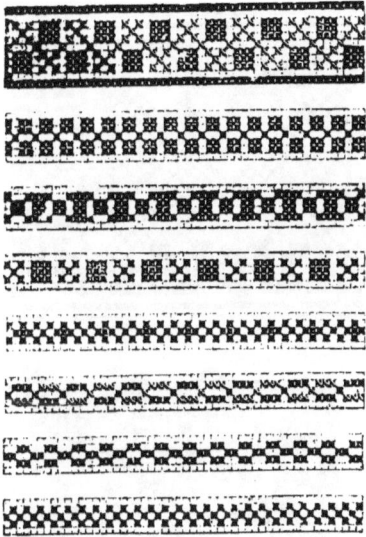

Fig. 168. BORDERS

Fig. 169

■ Black, 1st, (Darkest), ▨ 2d, ▨ 3d, (Lightest) Olive Green ▨ 1st (Darkest), ▨ 2d, ▨ 3d, (Lightest) Bronze. ✗ 1st, (Darkest), ✗ 3d ▨ 3d, (Lightest) Pink. ▢ Yellow.

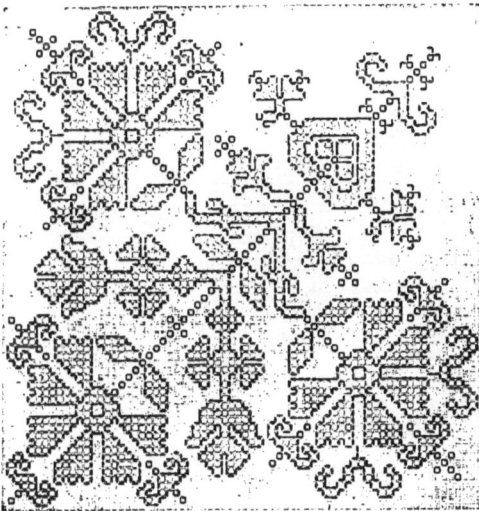

Fig. 170

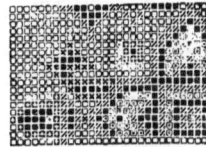

■ Dark Blue ✗ Green ▨ White ▨ Bronze ▨ Dark Green
DETAIL of Fig. 171

Fig. 171. NEWSPAPER HOLDER

Fig 172

Fig. 173. HANDKERCHIEF SACHEL

■ Dark Brown. ☐ Light Brown. ■ 1st, (Darkest). ▨ 2d. ▨ 3d, (Lightest) Green. ▨ Pink. ☐ Light Pink. ▨ Dark Blue.
☐ Light Blue. ▨ Gold)

DETAIL of Fig. 173

Fig. 174

Fig. 175

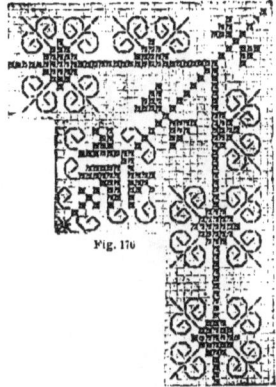

Fig. 176

Fig. 177

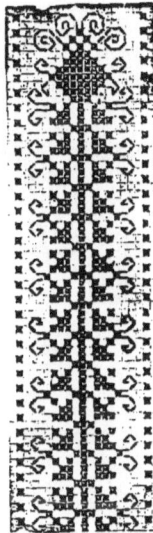

Fig. 178

Fig. 179

Fig. 180

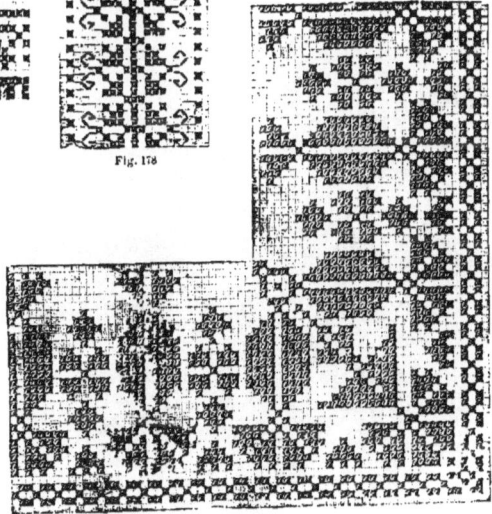

Fig. 181

Fig. 182

Fig. 183

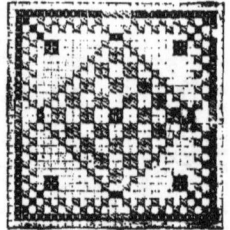

Fig. 184

■ Red. ▨ Green. ▨ Dark Yellow. ▨ Light Yellow. ☐ Old Blue. ▨ Ground.

DETAIL of Fig. 186

Fig. 185

Fig. 186. CHAIR-BACK

Fig. 187

Fig. 188

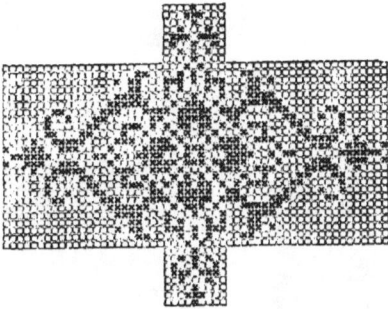

DETAIL of Fig. 189

DETAIL of Fig. 189

Fig. 189. CANVAS RUG

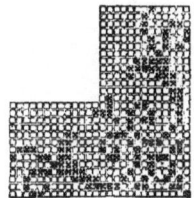

DETAIL of Fig. 189

Fig. 190. SOFA CUSHION

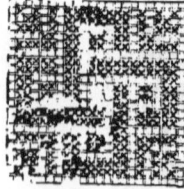

Fig. 191

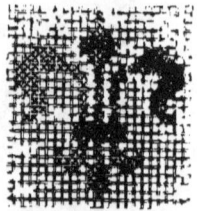

Fig. 192

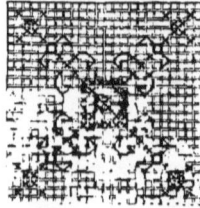

DETAIL of Fig. 190

Fig. 193

Fig. 195

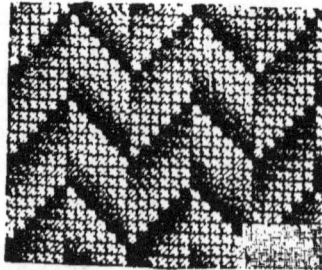

DETAIL of Fig. 195

Fig. 194

Fig. 196

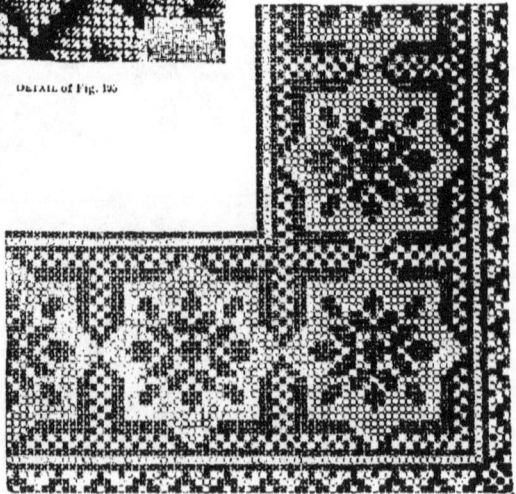

Fig. 197

www.ingramcontent.com/pod-product-compliance
Lightning Source LLC
Chambersburg PA
CBHW022023190326
41519CB00010B/1580